Bouzid Nedjimi

Physiological behavior of Sparte(Lygeum spartum L) to salinity

Bouzid Nedjimi

Physiological behavior of Sparte(Lygeum spartum L) to salinity

Salinity tolerance

ScienciaScripts

Imprint

Cover image: www.ingimage.com

This book is a translation from the original published under ISBN 978-3-8381-4223-4.

Publisher:
Sciencia Scripts
is a trademark of
Dodo Books Indian Ocean Ltd. and OmniScriptum S.R.L publishing group

120 High Road, East Finchley, London, N2 9ED, United Kingdom
Str. Armeneasca 28/1, office 1, Chisinau MD-2012, Republic of Moldova, Europe
Managing Directors: Ieva Konstantinova, Victoria Ursu
info@omniscriptum.com

Printed at: see last page
ISBN: 978-620-6-76376-5

Contents

Summary

Spartum (*Lygeum spartum* L.) is a plant of great ecological (against desertification) and economic (especially in the paper industry) value. It is one of the perennial grass species of the North African steppes. Its physiological behaviour in relation to NaCl (0, 50 and 100 mM) was studied at the germination phase and at the young stage in hydroponics and under controlled conditions. The germination rate of *L. spartum* seeds decreased by 25% at a salt concentration of 58mM NaCl. After two weeks of treatment, only the 100mM concentration reduced growth. Osmotic ($\Psi\pi$) and water ($\Psi\omega$) potentials, associated with a significant accumulation of sodium (Na^+) and chlorine (Cl^-), are strongly lowered, while turgidity ($\Psi\tau$) remained stable. Glycine betaine was accumulated in significant quantities as a function of the NaCl concentration applied, while the accumulation of sugars made only a small contribution to osmotic adjustment. The trials carried out in this experiment show that *L. spartum* is moderately tolerant to salinity. It could therefore be grown in salty steppe areas for fodder and industrial purposes.
Key words : *Lygeum spartum* L, salinity, ions, osmotic adjustment, glycine betaine, total soluble sugars.

1 INTRODUCTION

In Algeria, there is currently a gradual deterioration of the grass flora (especially Alfa and Spartum) in all natural arid and semi-arid ecosystems, hence the need to reintroduce perennial grasses to rehabilitate steppe rangelands (**Kadi-Hanifi**, 2003). It was in this context that we chose *Lygeum spartum* L., a perennial grass that is important in pastoral terms, especially as it is seriously threatened by animal pressure. All the studies carried out on this species agree that spartum is generally considered to be moderately tolerant of salt (Nedjimi, 2009; Nedjimi et al., 2010). Given these facts, any action to restore the ecological balance of degraded environments must be preceded by a better understanding of the ecophysiological requirements of the material to be used.

Salinity is one of the environmental factors that has a critical influence on seed germination in halophytes and glycoplytes (Ashraf and Foolad, 2005). Numerous studies have reported that plant seeds do not germinate when exposed to high levels of salinity (Meot-Duros and Magné, 2008). However, when placed in distilled water, seeds that had not germinated at high concentrations achieved germination rates equivalent to those of the no-salt control. This indicates that osmotic germination is reversible (Abdul Jaleel et al, 2007). In addition, the seeds of many plants show optimal germination in fresh water, indicating a need to reduce soil salinity, which manifests itself in the field after rainfall (Khan and Ungar, 1999).

Salinity reduces plant growth by altering the water and ion balance in tissues (Türkan and Demiral, 2009). In leaves, this phenomenon is associated with a drop in turgidity, following a reduction in the water potential gradient between the plant and the environment (Martinez-Ballesta et al., 2006). Compartmentalisation of ions between organs (aerial part/root part), tissues (epidermis/mesophyll), or even between cellular compartments
(vacuole/cytoplasm) is one of the mechanisms for adaptation to salt stress (Parida and Das, 2005). Generally, in tolerant plants, unlike sensitive plants or glycophytes, sodium (Na^+) is well compartmentalised in the vacuole (Tester and Davenport, 2003). Osmotic adjustment involves the accumulation mineral ions (Cl^-, Na^+) and/or organic solutes such as proline, glycine-beta'i'ne, sugars and organic acids (Ashraf and Harris, 2004). In many species, including grasses, the accumulation of soluble sugars is one of the manifestations of water and salt stress (Dubey and Singh, 1999; Karimi et al., 2005).

With increasing salinity in the environment, salinity-tolerant plants achieve osmotic adjustment by concentrating salts in their tissues, where they are used as *osmoticums*, but the quantities that need to be accumulated quickly become toxic. One adaptation strategy is therefore to synthesise *osmoprotectants* and accumulate them in the cytoplasm to counterbalance the effect of ion accumulation in the vacuole (Ashraf and Foolad, 2007).

Glycine betaine is a quaternary ammonium compound (QACs) that functions as an effective compatible osmolyte in plants subjected to salt stress (Raza et al., 2007). This compound, which has the particularity of being methylated, is derived either from choline (choline-O-phosphate) or from amino acids (proline-betaine, ß-alanine-betaï'ne) through the action of betaï'ne aldehyde dehydrogenase (BADH) (Yin et al., 2002). It is thought to effectively protect the components essential for photosynthesis, such as Rubisco and the photosystem II complex (PSII), thereby maintaining photosynthetic activity (Demiral and Türkan, 2009). It also acts as a water scavenger, stabilising enzyme structures and activity and maintaining the integrity of cytoplasmic membranes (Saneoka et al., 2001).

Soluble sugars are thought to play a major role in osmotic adjustment, and their contribution to lowering the osmotic potential under conditions of salt stress has been demonstrated by several authors (Fernandes et al., 2004). Sugar synthesis is stimulated by salt stress in many species, either by blocking glycolysis or by starch hydrolysis (Rathinasabapathi, 2000).

Lygeum spartum L. (Poaceae) is a spontaneous, perennial species found in arid and semi-arid Mediterranean regions (Nedjimi, 2016 b). With a fairly large above-ground and root biomass, it is an effective and relatively inexpensive tool for erosion and desertification and rehabilitating degraded land (Pugnaire and Haase, 1996). In industry, its interest lies in the use of its leaves in the manufacture of paper pulp (Djabeur et al., 2008). These characteristics make *L. spartum* an excellent species for rehabilitating degraded pastoral areas. However, few studies to date have assessed the salinity tolerance of this species (Nedjimi, 2009; Nedjimi et al., 2010; Nedjimi, 2011; 2013). In this context, an experiment was undertaken with *L. spartum* seedlings under hydroponic conditions to evaluate the effect of different NaCl concentrations on growth, mineral nutrition and osmotic adjustment in *L. spartum*, at an early stage of its development.

2 ECOPHYSIOLOGY OF PLANTS IN A SALINE ENVIRONMENT

Soil salinity, often linked to water scarcity, is a major issue for plant development in dry regions (Hanana et al., 2011). According to the FAO (2005), soluble salts have an impact on 7% of the world's agricultural land (920 million hectares). In addition, around 25% of irrigated soils are affected by salinity (Keren, 2000). Every year, 10 million hectares of cultivated land are abandoned because of salinisation through irrigation with mineralised water (Szabolcs, 1994). Climatic conditions and human activities contribute to the salinisation of soils in these areas. Intense evaporation from the air and low rainfall cause a water deficit, encouraging the accumulation of soluble salts in the soil. These can reach a point where the land is unsuitable for growing crops. (Qadir et al., 2001). The presence of soluble salts in a soil at a specific concentration influences the plant's physiological processes and restricts its production (Maggio et al., 2004).

Given that a plant's yield is regulated by the interaction between its genetic potential and its natural environment, saline soils could be developed. On the one hand, we can act on the soil itself by removing the excessive quantity of soluble salts it contains (Kotzer, 2005). This process requires adequate water availability and considerable expense to install a drainage system to remove the soluble salts from the profile. On the other hand, saline soils can be developed by growing plants that are resistant to saline conditions and of agricultural interest (Masters et al., 2007).

In the context of salt stress, plants face a twofold difficulty. Firstly, the presence of soluble salts reduces the soil's water potential and hampers its water supply. Secondly, the penetration of salts into tissues jeopardises normal cell function (Mahajan and Tuteja, 2005). Faced with this problem, not all plants are the same. Some species, known as glycophytes, cannot withstand high levels of soluble salts. In contrast, halophytes have developed physiological responses to guarantee their water supply while maintaining their metabolism in a highly saline environment (Malcolm et al., 2003).

Identifying salinity-resistant species and incorporating them into highly saline soils is a promising strategy for restoring and developing these marginal areas, which are unfavourable for most crops intended directly for human consumption (Weber et al., 2007).

1.1. Notion of salinity tolerance

The resistance of plants to salinity refers to their ability to survive in an environment of soluble salts without hindering their growth and development.

(Rathinasabapathi, 2000). However, the reaction of plant species to soluble salts varies according to the species in question, its variety, the content and type of salts, growing conditions and the growth stage of the plant (Djanaguiraman et al., 2006).

A plant grown in a saline environment is confronted with excessive diffusion of certain ions into its tissues, which may be eliminated or grouped together by various plant organs (such as tissues, cells or cellular compartments) (Chen et al., 2002). When the surrounding environment has a high salt content, the plant first undergoes an osmotic shock that cancels out the gradient in water potential between its internal and external environments, which can lead to significant tissue dehydration. In response, this tissue dehydration leads to an increase in the cell's water potential thanks to an increase in vacuolar content, which is amplified in plants by the closure of the stomata. According to Verslues et al (2006), this is the first passive stage in plant resistance to salinity. The second phase involves dynamic processes that progressively reduce the plant's osmotic potential, leading to a new stationary condition known as osmoregulation or osmotic adjustment (Munns, 2005).

At the overall plant level, NaCl ions, frequently present in saline solutions, penetrate through the roots and are transported by the xylemic sap to the stems and leaves. They are then either retained (includer plants) or transported by the phloem sap to the roots (excluder plants). Chloride and sodium ions enter the root cells, then pass through the root cortex to the stele (symplasmic pathway), where they are released into the xylem along with other minerals (Chen et al. 2002). At cell level, taking into account that Na^+ and Cl^- enter via non-specific transporters or ion channels (James et al., 2006). Necrosis and cell death are determined by the fact that the intercellular space (or apoplasm) of the leaves is saturated by the salt transported by the xylem. (Munns, 2002). In this compartment, the salt concentrates and causes the release of intracellular water, leading to rapid dehydration of the cells. Two categories of behaviour are aimed preventing salt overabundance in the apoplast:

a. The *Includer* type

Minerals from the roots collect and accumulate in leaf cells (Figure 1), particularly in the vacuolar space. To date, the processes involved in the passage of sodium through the vacuolar membrane or tonoplast are not fully understood. However, the most commonly accepted assumption is that sodium enters against its electrochemical gradient. The proton gradient produced by the tonoplast proton pump would provide the energy required to transport this ion (Yamaguchi and Blumwald, 2005). In this way, the vacuole would be stacked in Na^+ by means of a sodium-proton antiport Na $/H^{++}$ which would be maintained thanks to the accelerated operation of the proton pumps (Popova and Golldack, 2007).

It seems that harmful ions such as Na^+ and Cl^- are eliminated from sensitive sites in the cell and locked up in the vacuole, which can make up 90% of the cell volume, where they act as osmoticums (Tyerman and Skerrett, 1999). The accumulation ions in the vacuole contributes simultaneously to adjusting its osmotic potential and detoxifying the cytoplasm, where cellular metabolic processes take place (Koyro, 2006).

b. The *Excluder* type

Minerals transported to the leaves and not retained there are returned to the roots via the phloem (Figure 1). Here, it is the root cells that guarantee the protection of the aerial parts, by restricting the volume of sodium transported by the xylem and/or expelling it to outside environment (Sobrado, 2004).

This retention could be due to a blockage of toxic elements within the underground parts, or an exclusion at the outer base of the roots. Thanks to this process, the concentration of elements in these plants changes little between saline and non-saline soils (Bethoumieu et al., 2003). For many halophytes, such as Amaranthaceae ex. Chenopodiaceae, the ability to withstand salt stress is linked to the presence of specialised organs such as vesicular hairs (trichomes) and excretory glands on the leaves, which give them a more or less shiny, whitish appearance (Simon, 1997). These anatomical structures are specialised in the removal of Na^+ which reaches the leaves via the xylem and then the mesophyll. These trichomes store Na^+ , which is then released by their degeneration, thus preventing the toxic accumulation of this cation in leaf tissues. Generally speaking, the cellular systems responsible for ion transport in plants are still poorly characterised at genetic or molecular level. For this reason, the processes that regulate the circulation and distribution of NaCl at the level of the whole plant or at the cellular level are still poorly understood (Munns, 2005).

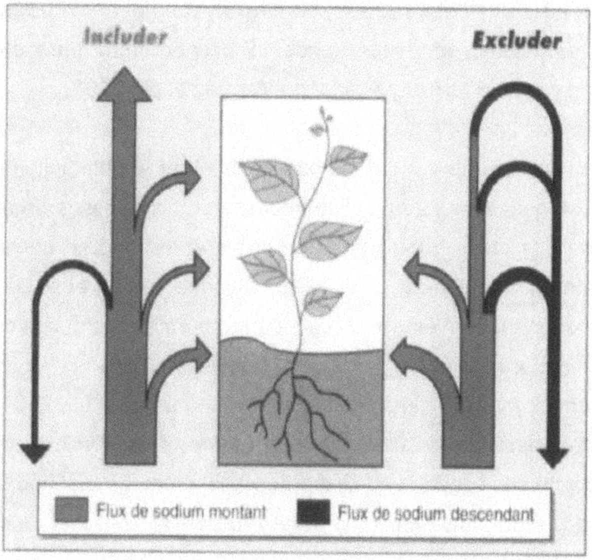

Figure 1: Diagram of sodium circulation in *includer* and *excluder* plants. In *includer* plants, sodium flows are mainly upwards (in pink) and the salt is accumulated in the aerial parts. In *excluder* plants, most of the sodium transported to the leaves is re-exported to the roots via the phloem (blue). The relative intensities of the flows are symbolised by the width of the lines (Levigneron et al., 1995).

1.2. Effects of salinity on plants

Salt stress occurs when the concentration of salts dissolved in the soil solution leads to a reduction in vegetative productivity. The tolerance threshold in terms of the concentration of soluble salts in the soil that plants can withstand without compromising their metabolism varies according to genus, species and variety (Munns, 2005).

Salinity can result in four forms of stress that soluble salts induce in plants: hydric, ionic, nutritional and oxidative stress.

a. *Water stress*

A significant concentration of soluble salts in the soil generates significant osmotic pressure in the root zone, which reduces the availability of soil water to the plant. This is known as physiological drought (Maricle et al., 2007). To cope with this osmotic pressure, the plant must perform osmotic regulation to ensure that the water potential of its cells remains lower than that of its extracellular environment and the soil. This process enables the plant to continue absorbing water from the soil and to conserve water inside its cells and support turgidity (Lee et al., 2005). When osmotic equilibrium is not achieved, water tends to escape from the cells, leading to dehydration and reduced turgidity (Redondo- Gomez et al., 2006).

b. Ionic stress

Metabolic activity is disrupted when the accumulation of soluble salts in tissues leads to ion toxicity (Maggio et al., 2007). The intensity of this stress is determined by the level of permeability of the membranes to ions and the degree of toxicity of these ions for the plant in question (Munns, 2005). Chlorine can be absorbed by the roots and concentrated in the leaves, which can lead to burning of the tips or edges of the leaves and premature yellowing. Chemical analyses of the soil or leaves can be carried out to check for possible chlorite toxicity (Gebauera et al., 2004). Sodium can also cause direct damage via its absorption by the roots. Typical signs of sodium toxicity leaf scorch, desiccation and tissue degradation at the leaf tips, as opposed to chlorine-induced symptoms which generally occur at the leaf tips (Tester and Davenport, 2003).

c. Nutritional stress

The salinity of the soil solution creates competition between ions at root level. This process leads to a lack of certain nutrients, as sodium competes with potassium and calcium via non-selective membrane transporters. Similarly, chlorine and sulphates compete with nitrates and phosphates (Jin et al., 2007). An overabundance of sodium and chlorine amplifies the permeability of the membrane wall, accelerating the outward diffusion of electrolytes and reducing membrane selectivity (Meychik et al., 2005). In extreme situations, an overabundance of Na_+ can cause damage to membranes, notably through the excessive substitution of divalent Ca^{2+} ions (Renault, 2005). So, to preserve essential processes in conditions of high salinity, it is crucial that the plant is resistant to dehydration and that ionic flows within tissues and cells are effectively regulated (Munns, 2005).

d. Oxidative stress

In addition to water and ion imbalances, salt stress leads to disruption of photosynthetic and metabolic processes via a secondary stress known as oxidative stress, resulting from the accumulation radical oxygen species (ROS), such as hydrogen (Ben Amor et al., 2006). Radical oxygen species are generally generated by normal cellular processes such as photorespiration, but their concentration increases under stress, leading to oxidative damage to membrane lipids, proteins and nucleic acids (Koca et al., 2007). It seems that the ability of plants eliminate

ROS or mitigate their harmful impacts is a crucial attribute of their tolerance to salt stress (Vranová et al., 2002). ROS-tolerant plants have the ability to prevent the generation of these peroxides, or to enhance the detoxification or restoration of damage caused by ROS (Bienert et al., 2006). ROS is suppressed by antioxidant compounds such as glutathione, thioredoxin, ascorbate and carotenoids, or by ROS-scavenging enzymes such as superoxide dismutase, catalase, glutathione peroxidases and peroxiredoxins (Hasegawa et al., 2000).

1.3. Physiological and metabolic effects

Firstly, salinity creates an osmotic phenomenon, which means that plants consume more energy to draw water from the soil. It can also have consequences for the plant due to the toxicity of a particular salt or on the assimilation of essential nutrients. These consequences disrupt plant functioning and inevitably lead to a reduction in production (Munns, 2005). The reduction in agricultural yield varies according to the plants' resistance to soluble salts, which is specific to each plant species (Rengasamy et al., 2003).

a. Germination

Salinity is one of the environmental factors that plays a crucial role in the germination of halophyte and glycophyte seeds (Ashraf and Foolad, 2005). Depending on the species, the inhibitory effect of salt stress on germination can be either osmotic (salt water uptake by the seed) or toxic (Abdul Jaleel et al., 2007).

When the salinity of the soil increases, the germination rate decreases, although the proportion of seeds that germinate can remain stable as long as the salinity does not exceed a certain limit, specific to the species and even the variety. Salt inhibits germination in two ways:

- By blocking water uptake by the embryo due to the high osmotic potential of the environment (Abdul Jaleel et al., 2007).
- By intoxicating the embryo with the harmful properties of certain ions (Naidoo and Kift, 2006).

b. Growth

The most common impact of abiotic stresses on plant physiology is reduced growth. This adaptive faculty is essential to the survival a plant subjected to abiotic stress, as emphasised by Ksouri et al. (2007). Indeed, this delay in development gives the plant the opportunity to save energy and resources to cope with the stress before the imbalance between the inside and outside of the organism reaches a point where the damage becomes irreversible (Misra and Gupta, 2005).

In the presence of salt stress, soil water availability decreases. For a plant to survive and develop, it needs to encourage the movement of water, caused by transpiration, against the osmotic, matrix and gravitational forces of the soil. To cope with the water deficit and maintain hydration and the turgidity of its tissues, the plant will encourage the absorption water by the roots by making an osmotic adjustment. This approach, used to maintain homeostasis in the face of salt stress, consumes a great deal of energy and resources, which is detrimental to growth (Verslues et al., 2006).

The idea of a biphasic growth response to salt stress was inspired by the importance of the time factor in the interaction between salt stress and growth (Munns, 2005). A rapid initial phase is attributed to the presence of high saline concentrations in the soil. This creates high osmotic pressure in the root environment, reducing the availability of soil water to the plant and causing cell dehydration. The second stage of the growth response is slower than the first. It results from the harmful effect of salt inside the plant, linked to its inability to prevent salt accumulating to harmful levels in its leaves. If the plant receives an overload of salts, this will lead to premature ageing of the leaves, which will reduce the production of assimilates produced by the plant and slow its growth. This is the second stage of the response, which clearly distinguishes species according to their ability to withstand dissolved salts .

The leaf mortality ratio is an essential element in the long-term survival of plants. If the continuous production of new leaves exceeds that of old leaves, there will be enough leaves capable of photosynthesis to sustain the life of the plant. However, if the mortality rate of old leaves exceeds that of new leaf production, the plant is doomed to perish (Munns, 2002).

Ions can have specific effects that are direct (toxicity) or indirect due to competitive uptake of ions, which influences the mineral nutrition of plants and alters their ionic status (Ashraf and Orooj, 2006).

c. Photosynthesis

Salinity affects plant productivity in a number of ways. On the one hand, it reduces the components of growth (Neocleous and Vasilakakis, 2007), and on the other, it directly influences photosynthesis at various levels. In general, salinity reduces net photosynthesis by reducing leaf area (Wang and Nil, 2000), by reducing the capacity of stomata to fix gases due to the drop in water potential (Neocleous and Vasilakakis, 2007), by reducing the chlorophyll present in leaves (M'rah et al., 2006), and by reducing photochemical activity (Redondo-Gomez et al., 2006). If the solar energy absorbed by the photosystems present in the leaf is no longer fully exploited through photosynthesis, toxic forms of oxygen can be formed, such as hydrogen peroxide (H_2O_2) and hydroxyl radicals (OH^-). These radicals are identical to those that drive the ageing process in animal and plant cells, inducing lipid peroxidation and the denaturation of proteins and DNA (Kukreja et al., 2005). It should be emphasised that, according to the argument put forward by Munns (2002), it is growth that is influenced before photosynthesis. The decrease in photosynthesis is mainly due to the closure of stomata and potentially the decrease in mesophyll (chlorophyll tissue) conductance, which are induced by the loss of turgidity and root signals (ABA) (Janicka- Russak and Klobus, 2007). In addition to the initial effects of water stress, if a toxic accumulation of Na^+ ions occurs in the aerial part, this can influence photochemistry and the enzymatic activity linked to the entire photosynthesis process (Broetto et al., 2007).

On a cellular level, salinity has an impact on the ultrastructure of chloroplasts (Parida et al., 2003; 2004 b), particularly that of grana (Hernandez et al. 1995). It leads to the disappearance of the chloroplast envelope, the emergence of lipid droplets and deterioration of the thylakoid membranes (Balnokin et al., 2004). The inhibition of oxygen release by these reactions can also lead to oxidative stress inside the cell through the creation of hydrogen peroxide (H_2O_2) (Bienert et al., 2006). These compounds are harmful and can cause damage to proteins, lipids and nucleic acids (Shevyakova et al., 2006).

In short, the reduction in plant productivity due to salinity is linked to a reduction in the capacity of the photosynthetic organ (leaf surface area) and a decline in the efficiency of photosynthetic processes associated with stomatal activity (Brown and Pezeshki, 2007).

d. Mineral nutrition

According to Malcolm et al (2003), the accumulation of ions, particularly Na^+ , appears to be the main adaptation strategy of halophytes in a saline environment. In fact, an increase in NaCl concentration in the environment results in a decrease in K_+ and Ca_+ levels in various halophytes such as Atriplex nummularia and Plantago coronopus [2]. On the other hand, Na_+ and Cl- levels increase rapidly as a function of external concentration (Ramos et al., 2004; Koyro, 2006). These authors also point to a strong correlation between salt tolerance, succulence and a low K $/Na^{++}$ ratio. The increase in Na^+ in plant tissues subjected to salinity is proportional to the drop in cellular osmotic potential, indicating that Na^+ plays a direct role in the osmotic regulation mechanism (Tester and Davenport, 2003).

1.4. Biochemical effects

Plant tolerance to salt stress has frequently been linked to biochemical changes within the plant. One of these changes concerns the altered expression of certain genes (Sreenivasulu et al., 2007). Some genes, such as those linked to photosynthesis, are under-expressed (Winicov, 1998), while others, such as those associated with the biosynthesis of osmoprotectants, are over-expressed (Ashraf, 2004). Adaptation is also expressed through the rapid synthesis of polypeptides (Yamaguchi et al., 2006). Certain proteins accumulated under salt stress conditions have been identified, such as those belonging to the HSP (Heat Shock Protein) and ELIP (Early Light Induced Protein) categories (Sun et al., 2002).Functioning at the cellular level is altered to support the plant's adaptation strategy, both at the cellular level and at the level of the plant as a whole. Salinity sensitivity is detected by the membranes exposed to the saline solution as well as throughout the organism thanks to the reduction in turgidity (Xiong et al., 2002). Transduction signals are then emitted. Together, these signals regulate the restoration of ionic and water balance in cells, repair and avoidance of lesions, and cell multiplication (Zhu, 2002). For example, Ca^{2+}-based signals will trigger the release of stress hormones such as ABA (Abscisic Acid) in response to a decrease in external osmotic potential (Xiong and Zhu, 2002). It should be noted that this hormone plays an important role in water retention and resistance to various stresses in plant species. Under salt stress, ABA acts as a signal transmitted by the roots via the vascular system to the leaves. This signal leads to a reduction in transpiration by closing the stomata (Khadri et al., 2007). It also water uptake by the roots and modifies their development (Hartung et al., 2002). However, salt reduces the concentration of other growth hormones such auxin, cytokinins and gibberilins (Zhu, 2001).

1.5. Molecular and genetic approaches

Plant adaptation to salinity requires physiological cellular modifications regulated by specific gene expression (Kuznetsov et al., 2007). It is possible that these specific genes are responsible for the production of proteins involved in the sequestration of Na^+ (H^+ -Atpases, NHX-type transporters) (Zörb et al., 2005; Debez et al., 2006), the synthesis of compatible osmolytes (proline, glycinebetaine, polyols) (Parida and Das, 2005) or the elimination of toxic compounds (ROS elimination enzymes) (Sreenivasulu et al., 2007). In order to gain a better understanding of how plants tolerate soluble salts, it is essential to gain a better understanding of the expression of these genes (Yamaguchi and Blumwald, 2005).

When there is a disturbance in the cellular process, as in the case of stress, it is reasonable to expect changes (induction or repression) in the expression of a certain group of genes. When looking for genes whose expression is stimulated, inhibited or modulated in a specific physiological context, two methods are usually used:

- The first technique detects proteins (end products of gene expression) whose concentration is modified by the treatment applied (Verslues et al., 2006);
- The second method focuses on changes in gene expression at an earlier stage, specifically at the level of messenger RNA (Seki et al., 2003). Thanks to the application of these various techniques, we have been able to compile a list of genes induced or proteins produced in response to salt stress. The sequence of the corresponding protein can be deduced from that of a gene or mRNA. The function of these proteins can be identified or suggested by comparing their sequences with those in the databases. Subsequently, biochemical work is carried out to confirm the presumed function. The 'protein' and 'mRNA' methods for examining the impact of salt stress on plants are based on a separate selection of gene products.

With regard to the 'protein' method, we are examining the protein content of control plants or plants exposed to a salt treatment, two-dimensional electrophoresis. This technique allows macromolecules to be separated according to their size and charge. It is possible to identify, collect and microsequence proteins that are only present in plants under stress or that are highly accumulated (Yoshida, 2002).mRNA method, we begin by establishing a bank in *Escherichia coli* bacteria, so that each bacterial colony (clone) has only one version of mRNA. Using molecular hybridization, it is possible to identify the mRNAs that are uniquely present or more abundant in plants under stress. These mRNA copies can then be recovered from the clones, in particular for sequencing.

Several proteins have been extracted from halophytes and glycophytes exposed to salt stress (Souza-Filho et al., 2003; Vera-Estrella et al., 2005). Osmotin, which was one of the first to be isolated from tobacco grown in the presence of sodium chloride and which appears during osmotic adjustment, can accumulate in considerable proportions, reaching up to 10% of total proteins (Levigneron et al., 1995). The intrinsic variety of proteins identified to date makes it difficult to establish a precise understanding of the essential molecular mechanisms required for plant development in a saline environment. The results obtained are often incomplete, but the proposals arising from the many studies currently underway in this field point to significant progress in the medium term. Genetic techniques are now being added to the various physiological, biochemical and molecular biological studies into the reaction of plants to salt stress. The response to salt stress is complex, and plant resistance must necessarily be a multi-genetic attribute (Sreenivasulu et al., 2007). To date, no criterion for response to salt stress has been studied that can be used individually as a basis for selection for resistance in the field. It is therefore extremely complex to attempt establish breeding models for salt-tolerant plant varieties. The lack of results for the creation of lines resistant to soluble salts stems from two factors. Firstly, the breeder has few reliable indicators for judging a variety's salt tolerance outside the field trial. the other hand, conducting field trials is made particularly complicated by the extreme variability of saline soils and the powerful interactions between salinity and other environmental factors. To overcome this challenge, it is essential have reliable markers, whether physiological or genetic, to help describe the resistance behaviour plants to soluble salts.

3 PRESENTATION OF THE SPECIES

Spartina (*Lygeum spartum* L.), common names: Albardine, Sennaq, is a typical perennial grass of the Mediterranean basin (Figure 2). It occupies a large area of the Algerian steppes, to around 02 million hectares (Le Houérou, 1995). *Lygeum spartum* is one of the xerophytic steppe grasses; it is very well known in the Mediterranean because of its use as a source of fibre (Nedjimi, 2016 b). It combines with all the other steppe species
Physionomically dominant species such as *Stipa tenacessima* (Alfa) and *Artimisia herba alba* (white wormwood)

Figure 2. Sparta (*Lygeum spartum* L.).

(Baba Aissa, 2000). This species can tolerate extreme conditions of aridity, salinity and high temperatures (Diaz and Honrubia, 1993).
Sparte is a perennial herbaceous tuft with solid culms 30-70 cm high, covered at the base with brown scales, and a creeping rhizome covered shiny scales. The leaves are junciform cylindrical, with narrow blades, 50 cm long and 1.5 mm in diameter, with pointed membranous ligules 05-11 mm long. In the vegetative state, the leaves of esparto are generally confused with those of esparto; they can be distinguished by the fact that these leaves do not pull like those of esparto and by their marked ligules, which are reduced to a rim in the case of esparto (Ozenda, 1977). In esparto and sparte, the roots are fasciculated, but the only difference between the two species is that esparto roots grow centripetally (towards outside of the clump), whereas sparte

can send its roots to a depth of more than 1.5 m. In *L. spartum*, the inflorescence is very distinctive, reduced to a single terminal spikelet enclosed in a spathe with a lamellar tip, replacing the glumes, inside which are two flowers (or rarely three flowers) horned at the base to form a tube covered with long silky hairs. The glumes are missing, the glumellae are polynervate (9 veins), 20-30 mm long with silky hairs 10-15 mm long, fused in their lower halves to form a cylindrical tube (Cope and Nesbitt, 2002). The flowers are hermaphroditic, bearing 03 stamens with 15mm long anthers. A hairless ovary, 02 stylets fused into a single style ending in a single stigma. Flowering generally takes place April and May. The seeds ripen from June to July. The fruit is a medium-sized (8-9 mm long), fusiform caryopsis containing a small embryo, with a hard, low-lipid, starch-rich endosperm.

Lygeum spartum is native to a large part of the Mediterranean basin, in the north; Spain, Italy including Sicily, Sardinia and Malta, and in the south; Morocco, Algeria, Tunisia, Libya and Egypt, but it is absent from Greece, Turkey and most of the Middle East. Surprisingly, it has also been found in Kashmir but was probably introduced as an ornamental plant because this region is not characterised by a Mediterranean climate (Cope and Nesbitt, 2002). In Algeria, *L. spartum* occurs spontaneously in the cool arid and cold semi-arid bioclimatic zones, and is present mainly in western Algeria, as far as the Sahara (Baba Aissa, 2000). Groups of *L. spartum*, found mainly in the Wilayas of Naâma and Saida, are generally located on gypsum glacis and depressions, but sometimes with traces of sulphates and chlorides (Celles, 1975). Examination of the distribution of *L. spartum* has shown that it finds its ecological optimum between isohyets of 200 and 250 mm/year with an average minimum temperature of around +2°C. In its natural state, the cover of this species is between 30 and 40%, but can sometimes reach 80% on sandy deposits (Aidoud et al., 2006). It appears in gypseous and even salty environments with an electrical conductivity of between 02 and 30 dS/m (Nedjimi et al., 2013; Nedjimi, 2014).

Lygeum spartum is an important economic source in the arid zones of Algeria, Morocco and Tunisia, and even in Spain. The leaves, which are rich in fibres, flexible and resistant, have been used to make belts, baskets, shoes, hats, ropes, baskets and wine bottle stoppers since ancient times (Cope and Nesbitt, 2002). In North Africa, esparto and esparto grass are important fibre resources, yielding 0.2 to 0.7 tonnes per hectare for paper pulp. They also provide raw materials for handicrafts. Sparta leaves contain 56% fibre, or around 10% more than straw, so they have inherited demand as a replacement for cloth in paper manufacture. For this purpose they first used by the French. The leaves should be harvested before they are fully ripe; if, however, they are too young, they provide a semi-transparent, well-filled and dense paper with excellent inking qualities (F.A.O, 1992).

Lygeum spartum has a forage production of 0.3 to 0.4 UF/kg DM. The steppes with *L. spartum* are moderately productive, but they nevertheless constitute fairly good quality rangelands, their interest coming from their floristic diversity. The relatively high productivity (110 kg DM/ha/year) of annual and perennial species gives these types of rangeland a high pastoral production of 200 to 300 FMU/ha/year, allowing a stocking rate of 2 to 5 sheep/ha (Aidoud et al., 2006).

Lygeum spartum is considered a natural barrier to wind erosion and dune stranding. Thanks to its highly developed root system, this species is a good example of adaptation to drought, but also to salinity (Nedjimi, 20016 a), in fact its presence on salty ground is particularly the result of invagination or tripling of the endodermis at the level of these roots (Kribi- Belabed, 1993). The species occurs spontaneously on former mining sites contaminated by various heavy metals. Recent studies have highlighted the species' promising phytoremediation potential (Conesa et al., 2007; Nedjimi and Beladel, 2015).

4 MATERIALS AND METHODS

1.6. Germ test

L. spartum seeds were collected the Aïη Maâbed - Djelfa region (Algeria) (2° 39' E longitude, 34° 35' N latitude and 934 m altitude). After manual decortication of their hairs and bracts, the seeds were disinfected for 20 min in 70% ethanol, then 10 min in an 8% sodium hypochlorite solution, rinsed thoroughly with distilled water, then germinated, in batches of 100 seeds for each treatment, in boxes 15cm long by 07cm wide, lined with sterilised filter paper, at a rate of 25 seeds/box, i.e. 4 replicates per treatment (Debez et al., 2004). The filter paper was moistened initially and then every 24 hours with 5 ml of saline solution. The germination test was carried out in a culture chamber with controlled temperature and photoperiod. The temperature was set at 25°C ± 1°C, under an illumination of 400 µmol m^{-2} s^{-1} provided by a series of Mazdafluor fluorescent neon lights. The photoperiod was 16 hours of light and 8 hours of darkness. Relative humidity was 60% (day) and 80% (night). Germinated seeds were counted every 24 hours. The appearance of a radicle about 01 mm long was used as the germination criterion (Bajji et al., 1998). The experimental set-up used was a complete randomized block (Dagnelie, 1981). Each treatment was repeated 4 times, each repetition corresponding to a petri dish containing 25 seeds.

1.7. Hydroponics

After 07 days, seedlings between 02 and 03 cm in height, germinated in vermiculite, were placed in 15-litre tubs containing a continuously aerated nutrient solution of Hoagland and Arnon modified (Hoagland and Arnon, 1938), with pH maintained between 5.5 and 6, and containing: Ca(NO3)2 (2 mM), K2HPO4 (0.5 mM), MgSO4 (0.5 mM), H3BO3 (25 µM), MnSO4 (2 µM),
ZnSO4 (2 µM), CuSO4 (0.5 µM), (NH^Mθ7θ24 (0.5 µM), Fe-EDDHA [Fe-ethylendiamino-di(o-hydroxyphenylacetic) acid] (20 µM) (Figure 3). After 13 days of culture (when the plants were 20 days old) the plants were treated with 0, 50 and 100 mM NaCl. Each treatment was repeated five times and each replicate has six plants (30 plants per treatment). Fresh and dry matter, concentration of total soluble sugars, glycine betaine and ionic assay were measured after 15 days of treatments, when the plants were 35 days old.

Figure 3. *Lygeum spartum* seedlings grown in a hydroponic solution culture chamber by Hoagland and Arnon (1938).

The plants are separated into aerial and root parts. Each part was weighed to determine its fresh matter (FM). The dry matter (DM) was determined after drying the samples in an oven at 60°C for 48 hours. Ionic determination was carried out by atomic absorption spectrophotometry (Perkin Elmer Analyst 300) on the product of a nitro-perchloric attack in the presence of sulphuric acid (tri-acid method). Chlorine was determined using a Dionex D-100 chromatograph with an ionpac AS124-4 mm (10-32) column, using Chromeleon/Peaknet 6.40 chromatography software.

The hydric (Ψ_ω) and osmotic (Ψ_π) potentials are measured using a pressure chamber (Turner, 1988) and a Digital osmometer (Roebling, Berlin) respectively. The turgor potential (Ψ_T) is calculated as the difference between the water and osmotic potential ($\Psi_T = \Psi_\omega - \Psi_\pi$).

Glucose, fructose and sucrose were determined using a Dionex- D-600 chromatograph with a Carbopac PA 10 column. All data were recorded and processed using Chromeleon/Peaknet 6.40 chromatography software. The quantification of sugars was carried out by comparing the peak areas with those of the standards. Glycine betaine was measured using the method of Greive and Grattan (1983)

The experimental design used was a complete randomised block with 05 replicates for each treatment. The results were subjected to an analysis of variance (ANOVA), with *Tukey*'s test at the 5% threshold to identify homogeneous groups, using *SPSS* 7.5 software.

5 RESULTS

1.8. Germination

The results obtained show that the percentage of germination of *L. spartum* seeds obtained after 15 days varies according to the NaCl concentration of the medium (Figure 3). In the control medium, the germination percentage reached its maximum with a value of 55%. For a concentration of 50 mM NaCl (corresponding to an osmotic pressure of 0.23 Mpa), the germination rate was 41%. However, the addition of 100 mM NaCl to the medium (corresponding to an osmotic pressure of around 0.47 Mpa) produced a sharp drop in the germination rate to 22%. Analysis of variance using a classification criterion showed a highly significant difference between the different NaCl treatments used ($P < 0.01$). In addition, the *Tukey* test at the 5% threshold shows a significant difference between the control, which is part of group (a), and group (b), which corresponds to a concentration of 100 mM NaCl (Figure 4).

The rate of germination reduction is illustrated in Figure 5, which shows the following linear statistical relationship: $y = -33x + 55.83$, $r^2 = 0.98$.

According to this equation, the search for the admissible critical threshold that causes a 25% reduction in the germination rate (Mass, 1990) corresponds to a concentration of 58 mM NaCl.

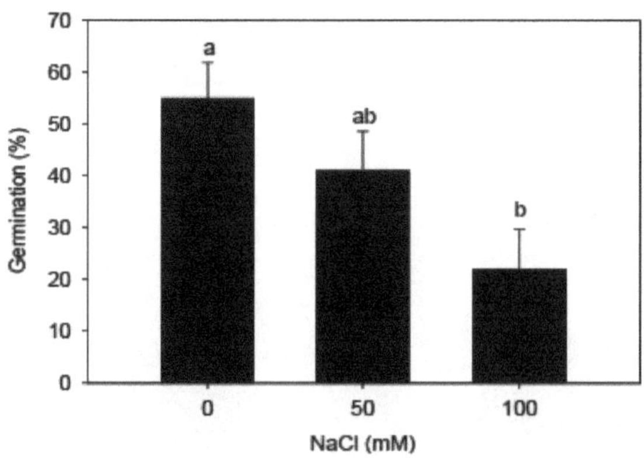

Figure 4: Effect of Nacl on the germination rate of *L. spartum* seeds. The bars represent the mean ± Standard deviation (n=4 replicates). The different letters above the bars indicate a significant difference at $P < 0.05$ according to *Tukey*'s test.

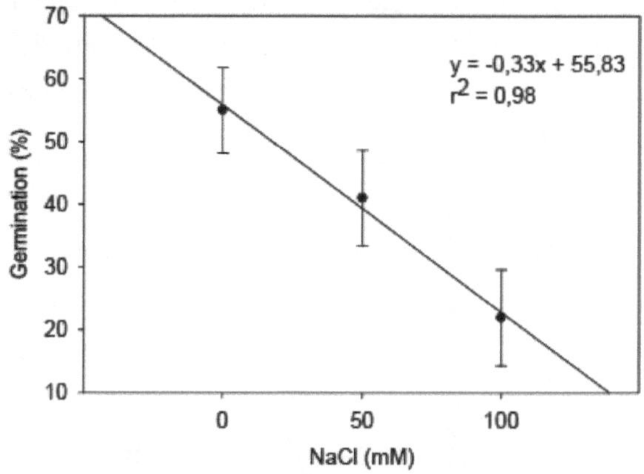

Figure 5. Statistical adjustment of the relationship between the germination rate of *L. spartum* seeds and the NaCl concentration of the medium.

The presence of sodium chloride causes germination to be delayed to a greater or lesser extent depending on the concentration of NaCl in the medium (Figure 6). The influence of NaCl on germination energy begins on the third day. On the fourth day, germination rates were 40, 28 and 14% for the 0, 50 and 100 mM NaCl treatments respectively while on the tenth day these rates reached their maximums with values ranging from 22 to 55%.

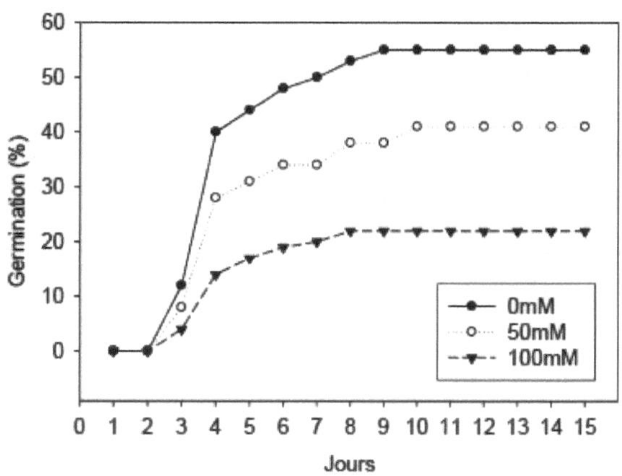

Figure 6. Germination rate of *L. spartum* seeds as a function of time and increasing doses of NaCl.

1.9. Dry matter

Dry matter production was not significantly affected by 50mM NaCl (Figure 7). In the presence of 100 mM NaCl, the reduction in growth compared with the control was around 33% for the above-ground part and 28% for the root part. According to the analysis of variance, the NaCl factor had a highly significant effect on dry matter production in the above-ground part ($P < 0.01$), it had no significant effect ($P > 0.05$) on dry matter production in the roots.

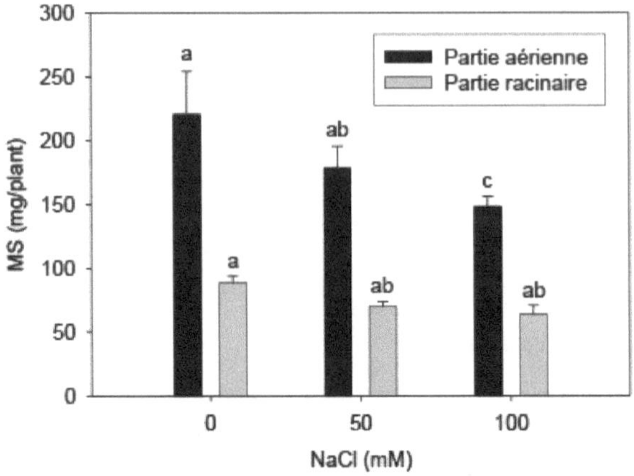

Figure 7. Effect of NaCl on dry matter (DM) of *L. spartum* plants grown in hydroponic solution. The vertical bars represent the mean ± SE (*n=5*). The different letters above the bars indicate a significant difference at $P < 0.05$ according to *Tukey*'s test.

1.10. Water status

Water (Ψω) and osmotic (Ψπ) potentials are lowered in the presence of NaCl; however, turgor remains stable, suggesting some level of osmotic adjustment (Figure 8). According to the analysis of variance, the NaCl factor exerts a very highly significant effect ($P < 0.01$) on water and osmotic potentials (Ψπ), while it has no significant effect ($P > 0.05$) on turgidity (Ψτ).

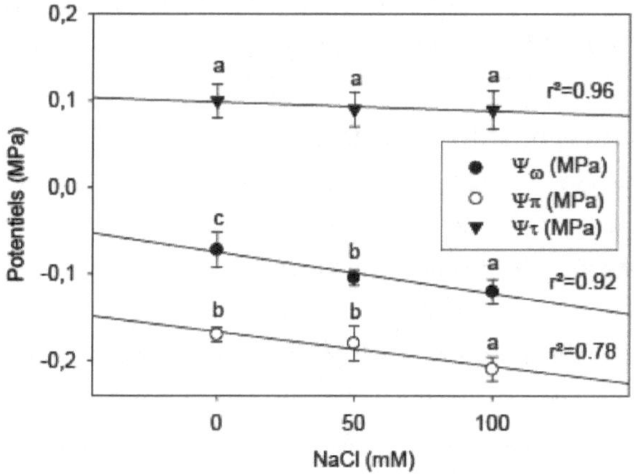

Figure 8. Effects of NaCl on water (Ψω), osmotic (Ψπ) and turgor (Ψπ) potential of *L. spartum* plants grown in hydroponic solution. Each value represents the mean ± SE (n=5). Different letters above the values indicate a significant difference at $P < 0.05$ according to *Tukey*'s test.

1.11. Mineral nutrition

Analysis of variance showed that increasing the concentration of NaCl in the medium significantly affected tissue levels of Na^+ , Cl^-, K^+ and Ca^{2+} in *L. spartum* ($P < 0.001$). At 100 mM NaCl, examination of the distribution of Na^+ and Cl^- accumulated within the plant showed that 65% (77.69mM) of Na^+ and 60% (81.13 mM) of Cl^- incorporated by the plant were compartmentalised in the aerial part, while the proportions exported to the root remained low, barely exceeding 35% (42.76mM) and 40% (55.54mM) respectively of total Na^+ and Cl^- (Table 1), so the plant's capacity to accumulate sodium and chlorine is preferentially localised in its aerial part. In addition, after 15 days of treatment, NaCl led to a reduction in K+ and Ca^{2+} levels in both parts of the plant (Table 1).

Table 1. Effects of NaCl on sodium (Na^+), potassium (K^+), calcium (Ca^{2+}) and chlorine (Cl^-) levels in the aerial and root parts of *L. spartum* plants grown in hydroponic solution. Values represent the mean ± SE (*n*=5). Different letters in the same column indicate a significant difference at $P < 0.05$.

NaCl (mM)	Na^+ (mM)	K^+ (mM)	Ca^{2+} (mM)	Cl^- (mM)
	Aerial part			
0	1.07 ± 0.11 d	26.94 ± 2.07 a	27.75 ± 3.06 a	31.65 ± 3.54 b
50	52.54 ± 1.78 b	13.72 ± 1.37 c	21.66 ± 2.31 b	70.63 ± 9.49 a
100	77.69 ± 3.23 a	12.02 ± 1.65 d	16.75 ± 1.25 d	81.13 ± 9.50 a
	Root part			
0	0.96 ± 0.07 c	32.92 ± 3.33 a	37.23 ± 2.06 a	21.02 ± 0.06 c

50	21.02 ± 2.11 b	17.66 ± 1.71 c	23.87 ± 2.37 bc	43.66 ± 0.31 ab
100	42.76 ± 4.17 a	13.54 ± 1.24 c	20.51 ± 0.95 c	55.54 ± 0.25 a

1.12. Accumulation of organic solutes

The values presented in Figure 9 show that the increase in salinity in the medium leads to an increase in glycine-betam content in the above-ground and root parts. In fact, in the presence 100 mM NaCl, there was a 50% and 60% increase in glycine betam compared with the control for the above-ground and root parts respectively. Statistically, NaCl concentrations significantly increased the glycine-betam content of above-ground and root tissues ($P < 0.001$).

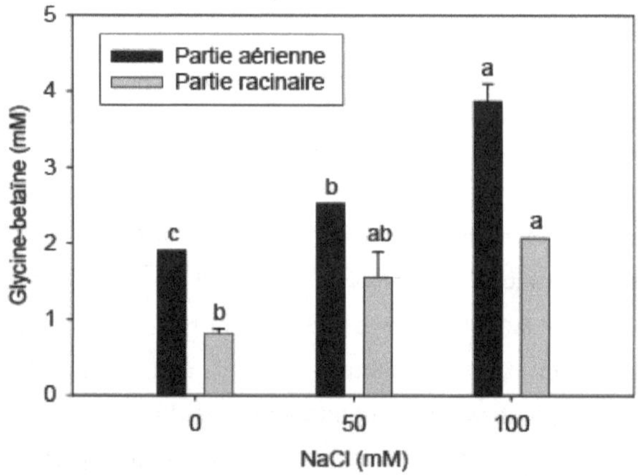

Table 9. Effects of NaCl on glycine-betaine content of *L. spartum* plants grown in hydroponic solution. The vertical bars represent the mean ± SE (*n=5*). The different letters above the bars indicate a significant difference at $P < 0.05$ according to *Tukey's* test.

Examination of Figure 10 shows that the soluble sugar content of leaves decreases slightly in the presence of NaCl. In the presence of the most stressful dose of NaCl, glucose, fructose and sucrose fell by 18, 25 and 27% respectively, compared with the control.

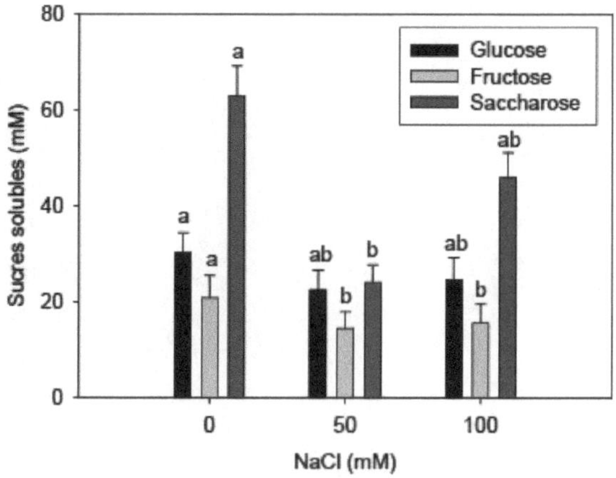

Table 10: Effects of NaCl on the soluble sugar content of *L. spartum* plants grown in hydroponic solution. The vertical bars represent the mean ± SE (*n=5*). The different letters above the bars indicate a significant difference at $P < 0.05$ according to *Tukey*'s test.

6 DISCUSSION

The research undertaken in this experiment is aimed at understanding one of the parameters of the intra-specific tolerance mechanism relating to perceived salinity at macro-element level, glycine betaine and soluble sugars accumulated in the above-ground and root biomass of a local poaceae species.

The tests carried out show that high salt concentrations slow down the germination rate of seeds and reduce their germination capacity. These effects are all the more marked the higher the saline concentration. These results are in agreement with those of Khan and Gulzar (2003) who noted a reduction in the rate and speed of seed germination in *Sporobolus ioclados* (perennial grass) when NaCl was added to the culture medium.

The observed disturbances could be explained by a reduction in the osmotic potential of the medium induced by the presence of salt, preventing the seeds from imbibing and triggering the germination process (Debez et al., 2004). However, once placed in distilled water, seeds that had not germinated at high salt concentrations reached germination rates equivalent to those of the no-salt control. This observation suggests that osmotic inhibition of germination is reversible (Khan and Gulzar, 2003). In natural conditions, this feature is of great interest, as it enables the survival of species whose natural habitats have wide fluctuations in salinity and opens up the possibility of building up a bank of viable seeds that germinate when environmental conditions become more favourable (Meloni et al., 2008).

The increase in the concentration of NaCl in the imbibition solution caused the germination period of *L. spartum* seeds to be extended. According to Redondo-Gomez et al (2007), this delay can be explained by the time needed for the seeds to trigger the mechanisms enabling them adjust their osmotic pressure. From an agronomic point of view, the delay caused by salt is not a constraint on the final yield of the crop, but is rather the germination capacity that is more decisive (Ashraf and Foolad, 2005).

Above a certain threshold, in all plant species, the salinity of the environment leads to a reduction in biomass. However, the degree of growth inhibition depends on the genus, species, variety, stage of development of the plant and nature of the organ, the nature of the salt and the salt concentration (Djanaguiraman et al., 2006). At an early stage in its development, esparto (*L. spartum*) behaves like a moderately salinity-tolerant plant, since its growth is significantly affected above 50mM NaCl. Similar results have been reported by Gulzar and Khan (2006) working on other Poaceae *Aeluropus lagopoides* and *Sporobolus ioclados*. The depressive effect of salinity on plant growth may have two main, non-exclusive causes relating to difficulties in supplying water and nutrients, and to the toxicity of ions accumulated in excess in the plant (Tester and Davenport, 2003). Our results suggest that the reduction in dry matter production observed with the addition of NaCl to the culture medium is attributable to the unfavourable action of sodium accumulated in the tissues.

Water status shows that turgidity ($\Psi\tau$) remains stable with increasing NaCl concentration, due to a significant drop in osmotic potential ($\Psi\pi$) relative to water potential ($\Psi\omega$). This decrease is mainly associated with the accumulation of Na^+ and Cl^- (Munns, 2002). Increased turgidity under the effect of salt has also been reported in other grasses such as *Paspalum vaginatum* by Lee et al. (2005) and in *Arthrocnemum macrostachyum* by Khan et al. (2005).

From a nutritional point of view, the presence of sodium chloride in the medium leads to increased uptake of Na^+ and Cl^- by the plant, resulting in an increase in the levels of these elements to the detriment of $Ca^{(2+}$) and K^+ . Competition between these elements is well known and has already been reported in other grass species such as rice (Alam et al., 2002) and barley (Shabala et al., 2005). These results confirm the hypothesis that salinity-tolerant plants use salts to adjust the osmotic pressure in their cells. They therefore allow salts to rise in their aerial parts and store them in the vacuoles where they are used as *osmoticums* (Parida and Das, 2005). However, at certain thresholds, this salt accumulation is accompanied by an inhibition of biomass production. In the case of *L. spartum*, this inhibition may be the result of a deficiency in other nutrients such as $Ca^{2+}$and K^+ , which are essential for maintaining cell structure, photosynthetic processes and regulating protein synthesis (Heller et al., 1998).

Increasing the saline concentration in the culture medium resulted in increased accumulation of glycine betaine in young *L. spartum* plants. This result suggests that the plant uses this compound for osmotic adjustment (Ashraf and foolad, 2007). These results corroborate those obtained by Rahman et al (2002), Yang (2003) and Raza et al (2007), who noted an increase in glycine betaine content in rice, sorghum and wheat subjected to salt stress. The accumulation of certain organic solutes (known under the name of osmoprotectants) is a common metabolic adaptation found in various plants subjected to salt stress. These dissolved bodies protect proteins and cell membranes from damage caused by high concentrations of inorganic ions. These osmoprotectants also protect metabolic mechanisms against oxidative damage (Parida and Das, 2005). Osmotic adjustment has been identified as an important mechanism in plant tolerance to salinity. This mechanism is a result of the accumulation of compatible organic bodies (such as glycine-betaine and soluble sugars) (Ashraf and Foolad, 2005). This response to salt stress may enable leaf turgidity to be maintained at a positive value, despite a drop in the water potential of the environment. This maintenance of tissue turgidity sustains the plant's metabolic activity (photosynthesis in particular), supports root growth and delays leaf senescence (Ashraf and Harris, 2005). As with glycine betaine, the results show that in *L. spartum* the cytoplasmic accumulation of soluble sugars is insignificant, suggesting that these osmoticums are only partially involved in the osmotic balance between the two endocellular compartments.

7 CONCLUSION

The main objective of this work was to study the effect of NaCl on germination, growth, mineral nutrition and the accumulation of certain osmotecums in a native grass species of forage and industrial interest (*Lygeum spartum*). The experimental approach in the laboratory involved germination tests and hydroponic cultivation under controlled conditions, followed by chemical and biochemical analyses.

In *L. spartum*, the germination rate reached 75% for a concentration of 58mM NaCl. This threshold corresponds to an osmotic pressure of around 0.27Mpa, giving an osmotic potential (expressed in pF) of 3.67. This level of potential generated solely by osmotic pressure is close to the temporary wilting point (generally set at a pF of 3.70) of most cultivated plant species (Mass, 1990).

This work has shown that salt stress has little effect on dry matter production at moderate salinity levels. Indeeddry matter accumulation in young *L. spartum* plants is significantly affected above 50mM NaCl, which corresponds to an osmotic pressure of 0.23 Mpa.

In terms of mineral nutrition, increasing the concentration of NaCl in the growing medium leads to an increase in Na^{+} and Cl^{-}, and a decrease in $Ca^{(2+})$ and K^{+} in the above-ground and root parts.

This species shows good hydration of its tissues in the presence of salt. In general, the water status of the aerial part shows that turgidity remains stable with increasing salt concentration.

Glycine betaine levels are negatively correlated with dry matter production. This osmoticum is involved in the osmotic adjustment of plant cells in saline conditions. Conversely, in the case of soluble sugars, the presence of sodium chloride results in a decrease in the content of these compounds, suggesting that these osmoticums play a partial role in osmotic balance.

These results provide a better understanding of the response of this plant to salt stress. They provide the data needed to propose cultivation practices that will make the most of saline soils through crop production for fodder and industrial purposes.

BIBLIOGRAPHY

Abdul Jaleel C., Gopi R., Sankar B., Manivannan P., Kishorekumar A., Sridharan R., Panneerselvam R. (2007) Studies on germination, seedling vigour, lipid peroxidation and proline metabolism in *Catharanthus roseus* seedlings under salt stress. *South Afr. J. Bot.* 73: 190-195.

A'ídoud A., Le Floc'h E., Le Houérou HN. (2006) The arid steppes of North Africa. *Sécheresse* 17 (1-2): 19-30.

Alam S., Imamul Huq S.M., Kawai S., Islam A. (2002) Effects of applying calcium salts to coastal saline soils on growth and mineral nutrition of rice varieties. *J. Plant Nutr.* 25: 561-576.

Ashraf M., Harris P.J.C. (2004) Potential biochemical indicators of salinity tolerance in plants. *Plant Sci.* 166: 3-16.

Ashraf M., Foolad M.R. (2005) Pre-sowing seed treatment- A shotgun approach to improve germination, plant growth, and crop yield under saline and non-saline conditions. *Adv. Agron.* 88: 223-271.

Ashraf M., Orooj A. (2006) Salt stress effects on growth, ion accumulation and seed oil concentration in an arid zone traditional medicinal plant ajwain (*Trachyspermum ammic* [L.] Sprague). *J. Arid Environ.* 64: 209-220.

Baba Aissa F. (2000) *Encyclopédie des plantes utiles, flore d'Algérie et du Maghreb, substance végétale d'Afrique, d'Orient et d'Occident.* Ed. Librairie moderne. Algiers, 268p.

Bajji M., Kinet J-M., Lutts S. (1998) Salt stress effects on roots and leaves of Atriplex halimus L., and their corresponding callus cultures. *Plant Sci.* 137: 131-142.

Balnokin Y.V., Kurkova E.B., Myasoedov N.A., Lun'kov R.V., Shamsutdinov N.Z., Egorova E.A., Bukhov N.G. (2004) Structural and functional state of thylakoids in a halophyte *Suaeda altissima* before and after disturbance of salt-water balance by extremely high concentrations of NaCl. *Russ. J. Plant Physiol.* 51(6): 815-821.

Ben Amor N., Jimenez A., Megdiche W., Lundqvist M., Sevilla F., Abdelly C. (2006) Response of antioxidant systems to NaCl stress in the halophyte *Cakile maritima*. *Physiol. Plant.* 126: 446-457.

Bethoumieu P., Conéjéro G., Nublat A., Brackenbury W.J., Lambert C., Savio C., Uozumi N., Oiki S., Yamada K., Cellier F., Gosti F., Simonneau T., Essah P.A., Tester M., Very A.A., Sentenac H., Casse F. (2003) Functional analysis of AtHKT1 in *Arabidopsis* shows that Na^+ recirculation by the phloem is crucial for salt tolerance. *Embo J.* 22: 2004-2014.

Bienert G.P., Schjoerring J.K., Jahn T.P. (2006) Membrane transport of hydrogen peroxide. *Bioch. Biophy. Acta* 1758 : 994-1003.

Broetto F., Duarte H.M., Lüttge U. (2007) Responses of chlorophyll fluorescence parameters of the facultative halophyte and C3-CAM intermediate species Mesembryanthemum crystallinum to salinity and high irradiance stress. J. *Plant Physiol.* 164: 904-912.

Brown C.E., Pezeshki S.R. (2007) Threshold for recovery in the marsh halophyte *Spartina alterniflora* grown under the combined effects of salinity and soil drying. *J. Plant Physiol.* 164: 274-282.

Celles J.C. (1975) *Contribution à l'étude de la végétation des confins Saharo-Constantinois (Algérie)*. Thèse doct. Etat, Univ. Nice, 364p.

Chen S., Li J., Fritz E., Wang S., Hüttermann A. (2002) Sodium and chloride distribution in roots and transport in three poplar genotypes under increasing NaCl stress. *Forest Ecol. Manag*. 168: 217-230.

Conesa, H.M., Robinson, B.H., Schulin, R., Nowack, B., 2007. Growth of *Lygeum spartum* in acid mine tailings: response of plants developed from seedlings, rhizomes and at field conditions. *Environ. Poll.* 145, 700-707.

Cope T., Nesbitt M. (2002) *Lygeum spartum. Curtis's Bot. Magazine* 19(1): 35-39.

Dagnelie Ph. (1981) *Principes d'expérimentation*. Les presses Agronomiques de Gembloux Belgique, 182p.

Davenport R., Reid R.J., Smith F.A. (1997) Sodium-calcium interactions in two wheat species differing in salinity tolerance. *Plant Physiol.* 99: 323-327.

Debez A., Ben Hamed K., Grignon C., Abdelly C. (2004) Salinity effects on germination, growth, and seed production of the halophyte *Cakile maritima*. *Plant Soil* 262:179189.

Debez A., Saadaoui D., Ramani B., Ouerghi Z., Koyro H., Huchzermeyer B., Abdelly C. (2006) Leaf H+-ATPase activity and photosynthetic capacity of *Cakile maritima* under increasing salinity. *Environ. Exp. Bot.* 57: 285-295.

Demiral T., Türkan I. (2006) Exogenous glycinebetaine affects growth and proline accumulation and retards senescence in two rice cultivars under NaCl stress. *Environ. Exp. Bot.* 56:72-79.

Diaz G., Honrubia M. (1993) Respuestas de crecimiento del albardin (*Lygeum spartum* L.) a la inoculacion con hongos micorrycicos y a la fertilizacion fosforada. *Cryptog. Mycol.* 14 :117-125.

Djabeur A., Kaid-Harche M., Khelifi D. (2008) Proteins polymorphism of some populations of *Lygeum spartum* L. in Algeria. *American J. Agric. Biol. Sci.* 3 (1): 337-

341.
Djanaguiraman M., Sheeba J.A., Shanker A.K., Devi D.D., Bangarusamy U. (2006) Rice can acclimate to lethal level of salinity by pre-treatment with sublethal level of salinity through osmotic adjustment. *Plant Soil* 284: 363-373.
Dubey R.S., Singh A.K. (1999) Salinity induces accumulation of soluble sugars and alters the activity of sugars metabolising enzymes in rice plants. *Biol. Plant.* 42 (2): 233-239.
F. A.O. (1992) *Foresterie en zones arides; guide à l'intention des techniciens de terrain*. Cahier F.A.O. Conservation. Rome, n°20. 143p.
FAO (2005) Global network on integrated soil management for sustainable use of salt-affected soils. Rome, Italy: FAO Land and plant nutrition management service. (http://www.fao.org/ag/agl/agll/spush).
Fernandes F.M., Arrabaça M.C., Carvalho L.M.M (2004) Sucrose metabolism in *Lupinus albus* L. under salt stress. *Biol. Plant.* 48(2): 317-319.
Gebauera J., El-Siddigb K., Salih A.A., Eberta G. (2004) *Tamarindus indica* L. seedlings are moderately salt tolerant when exposed to NaCl-induced salinity. *Sci. Horti.* 103 :1-8.
Greive C.M., Grattan S.R. (1983) Rapid assay for determination of water soluble quaternary-amino compounds. *Plant Soil* 70:303-307.
Gulzar S., Khan M.A. (2006) Comparative salt tolerance of perennial grasses. In: Khan M.A., & D.J. Weber (Eds.). *Eco-physiology of high salinity tolerant plants*. Dordrecht, The Netherlands, Springer, pp. 239-253.
Hanana M., Hamrouni L., Cagnac O., Blumwald E. (2011) Cellular mechanisms and strategies of salinity (NaCl) tolerance in plants. *Environ. Rev.* 19: 121-140.
Hartung W., Sauter A., Hose E. (2002) Abscisic acid in the xylem: Where does it come from, where does it go to? *J. Exp. Bot.* 53:27-32.
Hasegawa P.M., Bressan R.A., Zhu J.K., Bohnert H.J. (2000) Plant cellular and molecular responses to high salinity. *Ann. Rev. Plant Biol. Mol. Biol.* 51: 463-499.
Heller R., Esnault R., Lance C. (1998) *Plant physiology. I. Nutrition*. Ed. Dunod, 323p.
Hernandez J.A., Olmos E., Corpas F.J., Sevilla F., del Rio L.A. (1995) Salt-induced oxidative stress in chloroplasts of pea plants. *Plant Sci.* 105: 151-167.
Hoagland D.R., Arnon D.I. (1938) The water culture method for growing plants without soil. *California Agriculture Experiment Station Circular,* 347:1-39.
James J.J., Alder N.N., Mühling K.H., Läuchli A.E., Shackel K.A., Donovan L.A.,

Richards J.H. (2006) High apoplastic solute concentrations in leaves alter water relations of the halophytic shrub, Sarcobatus vermiculatus. *J. Exp. Bot.* 57(1): 139147.

Janicka-Russak M., Klobus G. (2007) Modification of plasma membrane and vacuolar H+-ATPases in response to NaCl and ABA. *J. Plant Physiol.* 164: 295-302.

Jin Z.M., Wang C.H., Liu Z.P., Gong W.J. (2007) Physiological and ecological characters studies on *Aloe vera* under soil salinity and seawater irrigation. *Process Bioch.* 42: 710-714.

Kadi-Hanifi H. (2003) Diversité biologique et phytogéographique des formations à *Stipa tenacissima* L. de l'Algérie. Sécheresse14 (3) :169-179.

Karimi G., Ghorbanli M., Heidari H., Khavari Nejad R.A., Assareh M.H. (2005) The effects of NaCl on growth, water relations, osmolytes and ion content in *Kochia prostrata. Biol. Plant.* 49(2): 301-304.

Keren R. (2000) Salinity. In: Sumner M.E. (Ed.) Handbook of Soil Science. CRC Press, NY,USA, pp: 3-25.

Khadri M., Tejera N.A., Lluch C. (2007) Sodium chloride-ABA interaction in two common bean (Phaseolus vulgaris) cultivars differing in salinity tolerance. *Environ. Exp. Bot.* 60: 211-218.

Khan, M.A., Ungar I.A. (1999) Seed germination and recovery of *Triglochin maritima* from salt stress under different thermoperiods. *Great Basin Naturalist* 59:144-150.

Khan M.A., Gulzar S. (2003) Germination responses of *Sporobolus ioclados*: a saline desert grass. *J. Arid Environ.* 53: 387-394.

Khan M.A., Ungar I.A., Showalter A.M. (2005) Salt stimulation and tolerance in an intertidal stem-succulent halophyte. *J. Plant nutr.* 28:1365-1374.

Koca H., Bor M., Özdemir F., Türkan I. (2007) The effect of salt stress on lipid peroxidation, antioxidative enzymes and proline content of sesame cultivars. *Environ. Exp. Bot.* 60: 344-351.

Kotzer E. (2005) Artificial kidneys for the soil - solving the problem of salinization of the soil and underground water. *Desalination* 185 :71-77.

Koyro H.W. (2006) Effect of salinity on growth, photosynthesis, water relations and solute composition of the potential cash crop halophyte *Plantago coronopus* (L.) *Environ. Exp. Bot.* 56: 136-146.

Kribi-Belabed S. (1993) *Contribution à l'étude morphologique du système racinaire du sparte* (*Lygeum spartum* L.) *en relation avec la sécheresse et la salinité.* Magister thesis, 72p.

Ksouri R., Megdiche W., Debez A., Falleh H., Grignon C., Abdelly C. (2007) Salinity effects on polyphenol content and antioxidant activities in leaves of the halophyte *Cakile maritima*. *Plant Physiol. Bioch*. 45: 244-249.

Kukreja S., Nandwal A.S., Kumar N., Sharma S.K., Sharma S.K., Unvi V., Sharma P.K. (2005) Plant water status, H2O2 scavenging enzymes, ethylene evolution and membrane integrity of *Cicer arietinum* roots as affected by salinity. Biol. Plant. 49(2): 305-308.

Kuznetsov V., Shorina M., Aronova E., Stetsenko L., Rakitin V., Shevyakova N. (2007) NaCl and ethylene-dependent cadaverine accumulation and its possible protective role in the adaptation of the common ice plant to salt stress. *Plant Sci.* 172: 363-370.

Lee G., Carrow V., Duncan A. (2005) Growth and water relation responses to salinity stress in halophytic sea shore *Paspalum* ecotypes. *Sci. Horti.* 104 :221-236.

Levigneron A., Lopez F., Vansyuyt G., Berthomieu P., Fourcroy P., Casse-delbart f. (1995) Les plantes faces au stress salin. *Cahiers Agri.* 4 : 263-273.

Le Houérou H.N. (1995) Considérations biogeographiques sur les steppes arides du Nord de l'Afrique. *Sécheresse* 6: 167-182.

Maas, E.V. (1990) Crop salt tolerance. In: Tanji, K.K. (Ed.), *Agricultural Salinity Assessment and Management*. ASCE Manuals and Reports on Engineering N°. 71, ASCE, New York, pp: 262-304.

Maggio A., De Pascale S., Angelino G., Ruggiero C., Barbieri G. (2004) Physiological response of tomato to saline irrigation in long-term salinized soils. *Europ. J. Agron.* 21:149-159.

Maggio A., Raimondi G., Martino A., De Pascale S. (2007) Salt stress response in tomato beyond the salinity tolerance threshold. *Environ. Exp. Bot.* 59: 276-282.

Mahajan S., Tuteja N. (2005) Cold, salinity and drought stresses: An overview. *Arch. Biochem. Biophy*. 444: 139-158.

Malcolm C.V., Lindley V.A., O'Leary J.W., Runciman H.V., Barrett-Lennard E.G. (2003) Halophyte and glycophyte salt tolerance at germination and the establishment of halophyte shrubs in saline environments. *Plant Soil* 253: 171-185.

Maricle B.R., Cobos D.R., Campbell C.S. (2007) Biophysical and morphological leaf adaptations to drought and salinity in salt marsh grasses. *Environ. Exp. Bot.* 60: 458-467.

Martinez-Ballesta M.C., Silva C., Lopez-Berenguer C., Cabañero F.J., Carvajal M. (2006) Plant aquaporins: new perspectives on water and nutrient uptake in saline environment. *Plant Biol.* 8: 535-546.

Masters D.G., Benes S.E., Norman H.C. (2007) Biosaline agriculture for forage and livestock production. *Agri. Ecosys. Environ.* 119: 234-248.

Meloni D.A., Gulotta M.R., Martinez C.A. (2008) Salinity tolerance in *Schinopsis quebracho colorado*: Seed germination, growth, ion relations and metabolic responses. *J. Arid Environ. 72:* 1785-1792.

Meot-Duros L., Magné C. (2008) Effect of salinity and chemical factors on seed germination in the halophyte *Crithmum maritimum* L. *Plant Soil* 313:83-87.

Meychik N.R., Nikolaeva J.I., Yermakov I.P. (2005) Ion exchange properties of the root cell walls isolated from the halophyte plants (*Suaeda altissima* L.) grown under conditions of different salinity. *Plant Soil* 277:163-174.

Misra N., Gupta A.K. (2005) Effect of salt stress on proline metabolism in two high yielding genotypes of green gram. *Plant Sci.* 169: 331-339.

M'rah S., Ouerghi Z., Berthomieu C., Havaux M., Jungas C., Hajji M., Grignon C., Lachaâl M. (2006) Effects of NaCl on the growth, ion accumulation and photosynthetic parameters of *Thellungiella halophila*. *J. Plant Physiol.* 163: 10221031.

Munns R. (2002) Comparative physiology of salt and water stress. *Plant Cell Environ.* 25: 239-250.

Munns R. (2005) Genes and salt tolerance: bringing them together. New Phytol. 167: 645-663.

Naidoo G., Kift J. (2006) Responses of the saltmarsh rush *Juncus kraussii* to salinity and water logging. *Aquatic Bot.* 84: 217-225.

Nedjimi B. (2009) Salt tolerance strategies of *Lygeum spartum* L.: A new fodder crop for Algerian saline steppes. *Flora* 204: 747-754.

Nedjimi B., Daoud Y., Carvajal M., Martínez-Ballesta M.C. (2010) Improvement of the adaptation of *Lygeum spartum* L. to salinity under the presence of calcium. *Comm. Soil Sci. Plant Anal.* 41: 2301-2317.

Nedjimi B. (2011) Is salinity tolerance related to osmolytes accumulation in *Lygeum spartum* L. seedlings? *J. Saudi Soc. Agri. Sci.* 10:81-87

Nedjimi B. (2013) Effect of salinity and temperature on germination of *Lygeum spartum* L. (Poaceae). *Agri. Res.* 2: 340-345.

Nedjimi B., Guit B., Toumi M., Daoud Y. (2013) Water status of *Lygeum spartum* L. seedlings subjected to salt stress. *BioRessources* 3:1-5.

Nedjimi B. (2014) Physiological behaviour of sparte (*Lygeum spartum* L.) in a saline

environment. *Nature Tech. 11: 30-33.*
Nedjimi B., Beladel B. (2015) Chemical element contents in *Lygeum spartum* L. using instrumental neutron activation analysis. *Radioch. Acta* 103: 463-466.
Nedjimi B. (2016 a) Salinity tolerance: Growth, mineral nutrients, and roles of organic osmolytes, case of *Lygeum spartum* L., A review. In: Iqbal N., Nazar R., Khan N.A. (Eds.) Osmolytes and plants acclimation to changing environment: Emerging omics technologies. Springer, Dordrecht, pp: 27-35.
Nedjimi B. (2016 b) *Lygeum spartum* L.: a review of a candidate for West Mediterranean arid rangeland rehabilitation. *The Rangeland J.* 38: 493-499.
Neocleous D., Vasilakakis M. (2007) Effects of NaCl stress on red raspberry (*Rubus idaeus* L. Autumn Bliss). Sci. Horti. 112: 282-289.
Ozenda P. (1977) *Flore du Sahara*. Ed. CNRS, Paris, 476p.
Parida A.K., Das A.B., Mittra B. (2003) Effects of NaCl stress on the structure, pigment complex compsition and photosynthetic activity of mangrove *Bruguiera parviflora* chloroplasts. *Photosynthetica* 41: 191-200.
Parida A.K., Das A.B., Mittra B. (2004) Effects of salt on growth, ion accumulation photosynthesis and leaf anatomy of the mangrove, *Bruguiera parviflora*. *Trees Struct. Funct.* 18: 167-174.
Parida A.K., Das A.B. (2005) Salt tolerance and salinity effects on plants: A review. Ecotox. Environ. Safety 60: 324-349.Pugnaire F.I., Haase P. (1996) Comparative physiology and growth of two perennial tussock grass species in a semi-arid environment. *Ann. Bot.* 77: 81-86.
Popova O.V., Golldack D. (2007) In the halotolerant Lobularia maritima (Brassicaceae) salt adaptation correlates with activation of the vacuolar H+-ATPase and the vacuolar Na $/H^{++}$ antiporter. *J. Plant Physiol.* 164: 1278-1288.
Qadir M., Schubert S., Ghafoor A., Murtaza G. (2001) Amelioration strategies for sodic soil: A review. *Land Deg. Develop.* 12: 357-386.
Ramos, J., Lopez, M.J., Benlloch, M. (2004) Effect of NaCl and KCl salts on the growth and solute accumulation of the halophyte *Atriplex nummularia*. *Plant Soil* 259: 163168.
Rathinasabapathi B. (2000) Metabolic engineering for stress tolerance: Installing osmoprotectant synthesis pathways. *Ann. Bot.* 86: 709-716.
Rahman M.S., Miyake H., Takeoka Y. (2002) Effects of exogenous glycinebetaine on growth and ultra-structure of salt-stressed rice seedlings (*Oryza sativa* L.). *Plant Prod. Sci. 5:* 33-44.

Raza S.H., Athar H.R., Ashraf M., Hameed A. (2007) Glycinebetaine-induced modulation of antioxidant enzyme activities and ion accumulation in two wheat cultivars differing in salt tolerance. *Environ. Exp. Bot.* 60: 368-376.

Redondo-Gomez S., Wharmby C., Castillo J.M., Mateos-Naranjo E., Luque C.J., de Cires A., Luque T., Davy A.J., Figueroa M.E. (2006) Growth and photosynthetic responses to salinity in an extreme halophyte, *Sarcocornia fruticosa*. *Physiol. Plant.* 128: 116-124.

Redondo-Gomez S., Mateos-Naranjo E., Wharmby C., Luque C.J., Castillo J.M., Luque T., Mohamed M.F., Davy A.J., Figueroa M.E. (2007) Bracteoles affect germination and seedling establishment in a Mediterranean population of *Atriplex portulacoides. Aquatic Bot.* 86: 93-96.

Renault S. (2005) Response of red-osier dogwood (*Cornus stolonifera*) seedlings to sodium sulphate salinity: effects of supplemental calcium. *Pysiol Plant.*123: 75-81.

Rengasamy P., Chittleborough D., Helyar K. (2003) Root-zone constraints and plant based solutions for dryland salinity. *Plant Soil* 257: 249-260.

Saneoka H., Ishiguro S., Moghaieb R.E.A. (2001) Effect of salinity and abscisic acid on accumulation of glycinebetaine and betaine aldehyde dehydrogenase mRNA in Sorghum leaves (*Sorghum bicolor*). *J. Plant Physiol.* 158: 853-859.

Shabala S., Shabala L., Volkenburgh E.V., Newman I. (2005) Effect of divalent cations on ion fluxes and leaf photochemistry in salinized barley leaves. *J. Exp. Bot.* 56: 1369-1378.

Seki M., Kamei A., Yamaguchi-Shinozaki K., Shinozaki K. (2003) Molecular responses to drought, salinity and frost: common and different paths for plant protection. Curr. Opinion Biotech. 14: 194-199.

Shevyakova N.I., Rakitin V.Y., Stetsenko L.A., Aronova E.E., Kuznetsov V.V. (2006) Oxidative stress and fluctuations of free and conjugated polyamines in the halophyte Mesembryanthemum crystallinum L. under NaCl salinity. Plant Growth Regul. 50: 69-78.

Simon L.E. (1997) Morphology, ontogeny and adaptive value of glandular mucilagene surfaces in the genus *Chinopodium* L. (Chenopodeaceae). *C.R. Acad. Sci. Paris, Sciences de la Vie* 320: 615-620.

Sobrado M. A. (2004) Influence of external salinity on the osmolarlity of xylem sap, leaf tissue and leaf gland secretion of the mangrove *Laguncularia racemosa* (L.) Gaertn. *Trees* 18: 422-427.

Souza-Filho G.A., Ferreira B.S., Dias J.M., Queiroz K.S., Branco A.T., Bressan-Smith

R.E., Oliveira J.G., Garcia A.B. (2003) Accumulation of SALT protein in rice plants as a response to environmental stresses. *Plant Sci.*164 :623-628.
Sreenivasulu N., Sopory S.K., Kavi Kishor P.B. (2007) Deciphering the regulatory mechanisms of abiotic stress tolerance in plants by genomic approaches. *Gene* 388:1-13.
Sun W., Montagu M.V., Verbruggen N. (2002) Small heat shock proteins and stress tolerance in plants. *Bioch. Biophy. Acta* 1577: 1-9.
Szabolcs I. (1994) Soils and salinisation. In: Pessarak M. (Ed.) Plant and crop stress. New York-Basel, Hong Kong, pp: 3-11.
Tester M., Davenport R. (2003) Na^+ tolerance and Na^+ transport in higher plants. *Ann. Bot.* 91: 503-527.
Türkan I., Demiral T. (2009) Recent developments in understanding salinity tolerance. *Environ. Exp. Bot.* 67: 2-9.
Turner N.C. (1988) Measurement of plant water status by the pressure chamber technique. *Irrigation Sci.* 9: 289-308.
Tyerman S.D., Skerrett I.M. (1999) Root ion channels and salinity. *Sci. Horti.* 78: 175235.
Vera-Estrella R., Barkla B.J., Garcia-Ramirez L., Pantoja O. (2005) Salt stress in Thellungiella halophila activates Na^+ transport mechanisms required for salinity tolerance. *Plant Physiol.* 139: 1507-1517.
Verslues P.E., Agarwal M., Katiyar-Agarwal S., Zhu J., Zhu J.K. (2006) Methods and concepts in quantifying resistance to drought, salt and freezing, abiotic stresses that affect plant water status. *Plant J.* 45: 523-539.
Vranová E., Inzé D., Van Breusegem F. (2002). Signal transduction during oxidative stress. *J. Exp. Bot.* 53: 1227-1236.
Wang Y., Nil N. (2000) Changes in chlorophyll, ribulose biphosphate carboxylaseoxygenase, glycine betaine content, photosynthesis and transpiration in *Amaranthus tricolor* leaves during salt stress. *J. Hortic. Sci. Biotech.* 75: 623-627.
Weber D.J., Ansari R., Gul B., Khan M A. (2007) Potential of halophytes as source of edible oil. *J. Arid Environ.* 68: 315-321.
Winicov I. (1998) New molecular approaches to improving salt tolerance in crop plants. *Ann. Bot.* 82 :703-710.
Xiong L., Schumaker K.S., Zhu J.K. (2002) Cell signalling during cold, drought, and salt stress. *Plant Cell* 14: 165-183.
Xiong L., Zhu J.K. (2002) Molecular and genetic aspects of plant responses to

osmotic stress. *Plant cell Environ*. 25: 131-139

Yamaguchi T., Blumwald E. (2005) Developing salt-tolerant crop plants: challenges and opportunities. *Trends Plant Sci*. 10(12): 615-620.

Yamaguchi K., Takahashi Y., Berberich T., Imai A., Miyazaki A., Takahashi T., Michael A., Kusanoa T. (2006) The polyamine spermine protects against high salt stress in *Arabidopsis thaliana*. *FEBS Letters* 580: 6783-6788.

Yoshida K. (2002) Plant biotechnology - genetic engineering to enhance plant salt tolerance. J. Biosci. Bioeng. 94(6): 585-590.

Yang W.J., Rich P.J., Axtell J.D., Wood K.V., Bonham C.C., Ejeta G., Mickelbart M.V., Rhodes D. (2003) Genotypic variation for glycine betaine in *Sorghum*. *Crop Sci.* 43: 162-169.

Yin X., Zhao Y., Luo D., Zhang H. (2002) Isolating the promoter of a stress-induced gene encoding betaine aldehyde dehydrogenase from the halophyte *Atriplex centralasiatica* Iljin. *Bioch. Biophy. Acta* 1577: 452- 456.

Zhu J. K. (2001) Plant salt tolerance. *Trends Plant Sci.* 6: 66-71.

Zhu J.K. (2002) Salt and drought stress signal transduction in plants. *Ann. Rev. Plant Biol.* 53: 247-273.

Zörb C., Noll A., Karl S., Leib K., Yan F., Schubert S. (2005) Molecular characterization of Na /H^{++} antiporters (ZmNHX) of maize (*Zea mays* L.) and their expression under salt stress. *J. Plant Physiol.* 162: 55-66

Kolibri 360 GmbH
Ferdinand-Jühlke-Straße 7
99095 Erfurt, Deutschland
produktsicherheit@kolibri360.de

Druck:
CPI Druckdienstleistungen GmbH
im Auftrag der
Kolibri 360 GmbH
Ferdinand-Jühlke-Str. 7
99095 Erfurt